Backyard Birds

Published by Wildlife Education, Ltd.
12233 Thatcher Court, Poway, California 92064
contact us at: **1-800-477-5034**
e-mail us at: **animals@zoobooks.com**
visit us at: **www.zoobooks.com**

Concept developed by KIDesign. Scientific Consultant: Mark Rosenthal, Curator Emeritus, Lincoln Park Zoo, Chicago, IL. *Photos:* Cover, Purestock (Getty Images); Half-title Page Bird, Royalty Free (Getty Images); Title Page Bird, Royalty Free (Alamy Images); Pages 2-3, Eric & David Hosking (Corbis); Page 4 Top, Michael S. Quinton (National Geographic); Bottom, Tom Brakefield/Photodisc Blue (Getty Images); Page 5 Top, Tom Brakefield/Photodisc Blue (Getty Images); Middle, Diane MacDonald (Photographers Choice); Bottom, Jeremy Woodhouse/Photodisc Red (Getty Images); Page 8 Left, Adam Jones (The Image Bank); Right, Purestock (Getty Images); Bottom, Klaus Nigge (National Geographic); Page 9 Left, Joel Sartore (National Geographic); Top, Purestock (Getty Images); Bottom, Jeremy Woodhouse/Photodisc Green (Getty Images); Page 12 Left, Joel Sartore (National Geographic); Middle, Chase Swift (Corbis); Right, Jeremy Woodhouse/Photodisc Blue (Getty Images); Page 13 Top, Hal Beral (Corbis); Middle, Gary Carter (Corbis); Bottom, Purestock (Getty Images); Page 14 Top, Frank Cezus (Photographers Choice); Bottom Right, Hans Reinhard/zefa (Corbis); Bottom Left, Photos.com; Page 15 Top Left, Darrell Gulin (Corbis); Right, Gary W. Carter (Corbis); Bottom Left, altrendo nature (Altrendo); Page 21 all: Photos.com.

Copyright © 2007 hardbound edition by Wildlife Education, Ltd.
All rights reserved. No part of this book may be reproduced in any form without written permission from the publisher.
Printed in China

ISBN: 978-1-932396-46-1

Backyard Birds

A bird has wings to spread, water to drink, feathers to fluff, feet to splash,

Bb

The words "bird" and "bath" both begin with the letter "b." The word "birdbath" is made by putting the words "bird" and "bath" together. Can you think of some other "double-b" words and phrases, such as "big bucket," "blueberries," or "brown bear"?

Bubble bath has a nice ring to it, too.

and a tail to wiggle in a birdbath.

What's in Your

Birds live all over the world, so wherever your backyard might be, take a look out the window! You're likely to see a backyard bird or two.

A **Steller's jay** is easy to spot in Seattle, Washington.

Backyard birds can be your friends.

People in Brownsville, Texas see **roadrunners** all the time.

Backyard?

Loons can be found swimming on Lake Buffalo in Minnesota.

In Ogunquit, Maine, **seagulls** are common.

You might find a **mockingbird** in your backyard if you live in Ocala, Florida.

5

Where'd Everybody Go?

Even if you're birdwatching, you might not see birds right away. Remember: be **patient** (birds are busy, and might be off doing something else), and be **quiet** (some birds blend in with their surroundings, so you might hear them before you see them).

While you're waiting, keep your eyes open for **other signs** that birds are nearby.

Look for **feathers**. You can tell what kind of bird has been around.

Some birds might even leave cracked **seeds** and other leftovers behind.

illustrations by Sophie Kittredge

Different kinds of birds build different kinds of **nests**. What can you tell about these birds?

Backyard birds lay all kinds of **eggs**. Their size, color, and number vary depending on what kind of baby birds are inside!

Sometimes, birds leave **footprints**. You can see where they went!

high / low

Some bird evidence can be found up **high**, in trees. Some bird evidence can be found down **low**, on the ground. Where might you find the things shown on this page?

7

Birds of a Feather

Birds are famous for their feathers. People recognize birds by their feathers, and they're probably how birds recognize one another, too! Feathers are one of the things that make backyard birds so beautiful.

female birds

male birds

But did you know that the male and female of some bird species have different body shapes and feather colors? See if you can match the male and female **cardinal**, **oriole**, and **bluebird**.

Dark-eyed Junco

The plump gray junco
under the feeder
kicks up bits of snow
with its strong little feet.

It finds cracked corn
and sunflower seeds,
then soars with its cousins
into a junco tree.

Poems by Jill Barrie
Illustrations by Pamela Carroll

Ruby-throated Hummingbird

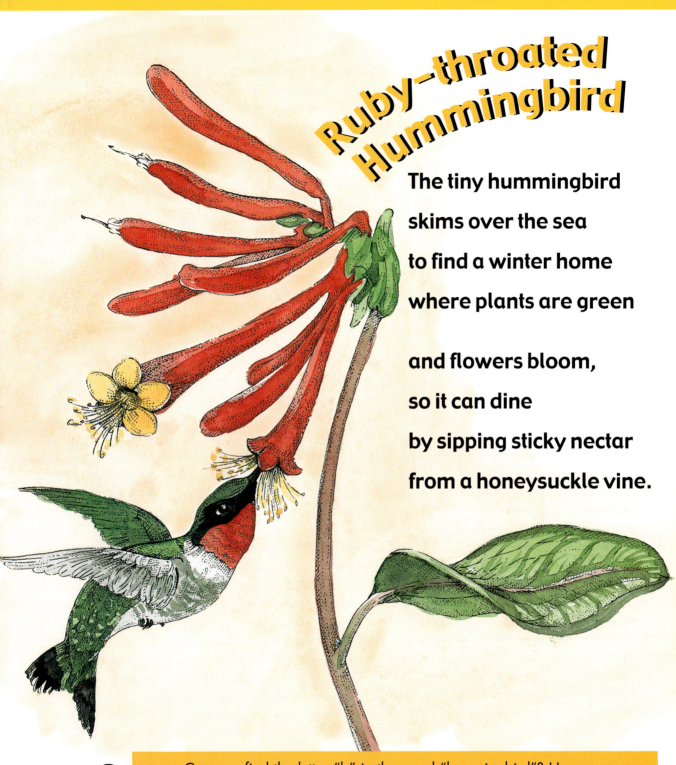

The tiny hummingbird
skims over the sea
to find a winter home
where plants are green

and flowers bloom,
so it can dine
by sipping sticky nectar
from a honeysuckle vine.

Bb

Can you find the letter "b" in the word "hummingbird"? How many "b's" can you find in the poem about the ruby-throated hummingbird? How about the other poem? Which poem has more "b's"?

Now can you find the letter "b" in the names of these backyard birds?
warbler, **robin**, **phoebe**, **bluebird**, **grosbeak**, **bunting**

What's for Lunch?

You can tell a lot about what a bird eats simply by looking at its beak and feet.

A **woodpecker's** special feet help it "hop" up and down the trunk of a tree. Its beak is strong enough to peck for insects.

A **heron's** feet aren't webbed, so it can walk quietly in water, and its long beak lets it grab fish without dunking its eyes underwater!

A **duck's** webbed feet are for swimming, and its round bill scoops underwater plants.

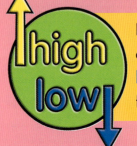

Birds fly differently, too! Some birds, like ducks, usually fly **low**, closer to the land and water. Other birds, like hawks, often fly **high**, over the treetops. Do you think a woodpecker would fly **high** or **low**? How about a hummingbird? Why?

B-b-bite-sized p-pieces?!?

A **hawk** has strong feet for grabbing prey and a sharp beak for ripping it into bite-sized pieces.

A **hummingbird** uses its long, thin beak to draw nectar from flowers. It hardly uses its feet, because it doesn't land very often!

A **cardinal's** short, sharp beak is perfect for cracking open seeds. Its feet are for perching on branches.

Bountiful Birds
by Rachel Young

Every winter, hungry birds dine at backyard feeders. And every winter, thousands of people, including kids, help scientists count the birds that visit those feeders.

David Bonter at Project FeederWatch collects the bird counts people send. Then he and other scientists use that information to figure out how many birds there are altogether in a given area, which types of birds are doing well, and which might need some help. Scientists could never count birds all over the country every winter on their own, says Bonter. "We rely on volunteers to be our eyes and ears in woods and backyards."

blue jay

Now that's what I call a mouthful!

You Can Be a Scientist, Too!

Would you and your family like to help scientists count birds? First, you'll need a feeder. David Bonter says the best place to hang a feeder is near trees, so birds will be hidden from predators—but not too close, or squirrels might jump from branches to help themselves to the birdseed. Fill the feeder with sunflower seeds, a food that many birds like. Then watch from a window and wait for the birds to arrive.

dark-eyed junco

downy woodpecker

black-capped chickadee

Chicka-Dee-Dee-Dee!

by Charnan Simon

Chicka-dee-dee-dee! Pip, the little black-capped chickadee, hopped busily under the privet hedge. Pip scratched in the frozen dirt, looking for insects. It was hard finding food in winter. Pip flitted over the snow to a nearby holly bush. He settled in to make a breakfast out of the bright red berries. But still—Pip was waiting, waiting . . .

illustrations by Kristin Kest

Ank! Ank! Ank! Jocko the white-breasted nuthatch swooped past Pip. Jocko flew to the top of the old maple tree. He began climbing his way headfirst down and around the trunk of the leafless tree. Jocko looked at the world upside-down! He found insects in the wood that other birds missed. Today, he also found an acorn he had stored last week, wedged under a loose slab of bark. Jocko used his sharp beak to hack, hack, hack that acorn to pieces. But still—Jocko was waiting, waiting . . .

Flash! Mr. Cardinal flitted to a high branch of the tall white pine. Nobody could miss his bright red feathers! Mrs. Cardinal was more private. She preferred to look for breakfast under the thick tangle of barberry bushes. Mr. and Mrs. Cardinal had nested in this yard for years. There were brush piles to hide in and good things to eat—sunflowers and coneflowers, crabapples, Juneberries, and tall wild grasses. But food was scarce in winter, and on this bright cold morning, Mr. and Mrs. Cardinal were waiting, waiting . . .

And here it came! The back door of the house opened, and the people came out. They scooped sunflower seeds into the bird feeder. They put suet into the suet basket. They poured fresh water into the wide, shallow bird bath.

Friendly little Pip flew right over. Thank you! he seemed to say as he began his good breakfast. Chicka-dee-dee-dee!

Bird Words — Bb

Zootles Fun Pages

These **backyard birds** all have at least one letter "b" in their names. The other letters have been filled in. Can you photocopy this page and write a "b" in each blank?

war__ler

gros__eak

humming__ird

__lack__ird

ro__in

__lue__ird

Here are some more "b" words you can find pictured in these photos:

The mother robin is feeding her b_____. They are very hungry.

The hummingbird has a very long b_____ for finding nectar in flowers.

The bluebird is perching on the b_____ of a tree.

Think you know the letter "b"? Are you ready for these **brain busters**?

Backyard Birds features the letter "b." While the uppercase B poses few problems, when kids are learning to write, the lowercase b can be a problem. Distinguishing between b, d, and even p can be daunting. Many children don't master the difference until first grade.

Positional concepts such as high/low, up/down, and first/last often take a back seat to learning color names or numbers, but they are part and parcel of the preschooler's cognitive development. Backyard Birds helps your child begin to think about positional words and how they are used.

Who's WHO?

Vertebrates
Birds
22 orders, representing more than 9,000 species!

Many "backyard birds" belong to the Order Passeriformes. This order consists mostly of songbirds and perching birds. More than 5,000 species—over half of all living birds—are Passeriformes.

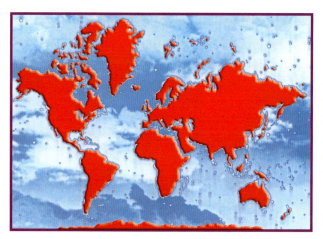

Where in the WORLD?

Wherever there's a backyard, there can be birds. Even in Antarctica, scientists often have penguins outside their research stations!

● **Birds**

Reading Resource

Every title in the *Zootles* series is designed to be used for fun and learning, and as a reading resource as well. The pages are written simply and address various stages of emerging literacy, and they encourage new readers to exercise their new skills at just the right level. Reading *Zootles* together will provide "together time" for you and your child—and reinforce vocabulary, comprehension, and early reading skills, too.

It's a ZOO out there!

Birds are wonderful to watch and easy to find. Just look out your window! For a preschooler, a sparrow eating a bread crust can be every bit as exciting as a flamingo sifting for shrimp, but if you're up for a family outing to a bird sanctuary, consider a place with water birds. Raptors or brightly colored songbirds can be hit-and-miss, but waterfowl tend to be there when you arrive—and usually in large numbers. Wherever you find birds to watch, think about the bird features discussed in this issue:
- What are the birds doing? Looking for food? Making a nest? Singing a song?
- Are the birds brightly colored to attract attention, or more dull for hiding?
- What do the birds' beaks and feet look like? What do you suppose they eat?
- Where do the birds like to hang out? High in a tree, on a wire, or a rooftop? Or low on the ground, or on the water?

What ELSE can we DO?

- **What's in Your Backyard? (p. 4-5)** helps us appreciate the diversity of bird life across the United States. What may be nothing more than a little brown bird to you might be considered exotic if it showed up on the other side of the Rockies!
  - **ZOOTLES TO-DO:** Toddlers probably aren't up to a dawn trip to the bird sanctuary with binoculars and a bird guide. With a digital camera, though, it's easy to snap a photo of something you see in your yard, then you and your child can try to match it to a picture in a bird book. Save the picture to a "life list" computer file, or print it out to glue into a journal.

- **Where'd Everybody Go? (p.6-7)** introduces an important technique used by field biologists. Not only do they spend time observing animals; they also take a close look at what their subjects have left behind: their homes, their food, and where they've been spending their time.
  - **ZOOTLES TO-DO:** Go on a nest hunt. In the spring, keep your eyes open for a bird gathering twigs and grasses, and try to track where it's headed. But it's much easier to spot nests in fall or winter after the leaves are down. Take a walk around the neighborhood with your head up, keeping an eye on the tree tops. Your child's vision may prove keener than yours!

- **Birds of a Feather (p. 8-9)** While the reds and yellows in feathers are formed by pigments, blues are created by the way the structure of the feathers reflects the light! If you hold a flashlight behind a blue feather, it will actually look brown.
  - **ZOOTLES TO-DO:** Wild bird feathers can be hard to find, and many shouldn't be picked up, but you can use colorful craft feathers to make your own feathered friend. Find a picture of a bird in a magazine or trace a picture from a book. Glue googley-eyes over the eyes. Staple or glue pipe-cleaner "feet" over the feet. Glue feathers on the rest. Up, up, and away!

- **Dark-eyed Junco** and **Ruby-throated Hummingbird (p. 10-11)** translate into verse what we see with our eyes.
  - **ZOOTLES TO-DO:** Use the poems as a model. Watch a backyard bird closely together. Then ask your child to dictate a poem, like this:
    Line 1. Describe your bird and say its name.
    Line 2. Tell exactly where the bird is.
    Line 3. Say what the bird is doing.
    Line 4. Say what part of its body it is using.
    (Of course, the poem will need an illustration before it goes up on the fridge.)

- **What's for Lunch? (p. 12-13)** Birds' beaks can peck, crack, rip, scoop, and tear. Their feet can perch, grab, climb, and wade. Every bird is perfectly adapted to its niche, its parts coming together to help it keep itself fed.
  - **ZOOTLES TO-DO:** Talk with your child about how your hands and mouths help you make and eat your food, too. Prepare and share a snack to prove it!

- **Bountiful Birds (p. 14-15)** describes a real form of scientific research where anybody can participate.
  - **ZOOTLES TO-DO:** You and your family can take part in Project FeederWatch or its sister project, The Great Backyard Bird Count. All you need to do is set up a feeder and count the birds that visit. For information, go to http://www.birds.cornell.edu/pfw/ or http://www.birdsource.org/gbbc/.

- **Chicka Dee Dee Dee (p.16-19)** gives a birds'-eye-view on the activity around a backyard bird feeder.
  - **ZOOTLES TO-DO:**
    - Feeding birds can be simple; just toss bread crusts out on the ground or rub some cheese on the rough bark of a low branch of a tree.
    - Or make a hanging birdfeeder by rinsing a plastic milk jug and cutting a hole in one side, about an inch from the bottom. Add seed and hang the feeder with a piece of string tied to the handle.
    - You can also make a bird cake to put in a suet basket. Melt a cup of vegetable shortening, suet, or lard in a microwaveable container. Add a cup of wild bird seed, or chopped raw peanuts, bits of cheese, and cornmeal. Spread your bird cake at the bottom of the feeder and let it harden.

Allie: An intrepid hedgehog

"Would you like to write a poem about me, Otto?"

Otto: An adventure-loving otter

Zootles Resource Corner